Common E

Best Guide On Everything You Need To Know About Common Eastern Frog Let As A Pet

Joseph Thomas

Table of Contents

CHAPTER ONE

COMMON EASTERN FROGLETS

If you stay in eastern Australia or Tasmania, you can listen a tiny little frog chatting away, seeking to appeal to a mate.

Common Eastern Frog lets are very small, only 1.Eight to a few cm lengthy, and are the most common and significant frog in south-Japanese Australia.

Common Eastern Frog lets are common backyard site visitors. They'll happily live in and round garden ponds, swimming pools, and ditches of water in suburban and urban regions. They are simply as commonplace in rural regions, residing close to farm dams, swamps, flooded grasslands, and just about everywhere there are pools of freshwater.

Eastern Frog lets devour small bugs inclusive of flies, mosquitoes, cockroaches, and spiders, millipedes and caterpillars.

While a few frogs best call while the climate is warm or hot, Common Eastern Frog lets want no unique conditions. They will call away all through the day or night, and all 12 months round. Their call is soft and easy, and hastily repeated.

Listen out for the adult males calling "crick, crick, crick, crick, crick" collectively. They sound a piece like crickets, except at a lower pitch. You may spot a Common Eastern Frog let calling from amongst the moist flowers on the water's facet, or within the center of the water delicately balanced on a plant.

If you turn over a wet rock, log or maybe a few leaf litters at the same time as gardening, don't be too amazed if you spot dozens of those little buddies. They love to cover wherein it's moist and darkish. You may additionally even trace their amplified calls lower back to a drain, in which moisture lingers and frogs experience secure and secure and could occasionally mate.

Your backyard frogs may be simpler to identify at night, when they can pass around within the open without drying out and without being too uncovered to predators.

Common Eastern Frog lets breed almost 12 months spherical, and a girl could have multiple take hold of each yr. Males call to draw a mate, and in case you're listening to them, take a look at your outside pond or nearest water source for little eggs or tadpoles. These frogs lay their eggs both singly, or in small clumps attached to underwater plant life. Sometimes their eggs roll around the bottom of the pond.

It takes about per week to 10 days for the eggs to hatch, after which the tadpoles can take everywhere from six weeks to 3 months to become person frogs. Common Eastern Frog lets vary substantially in shade, markings and length.

DESCRIPTION

THREATS

There are not any current threats to this present day frog populace but increasing urban development along the east-coast can be a threat inside the near future, but possible threats are urbanization and tourism, loss of temporary pools, and habitat fragmentation. Studies have shown that when habitat fragmentation the species disappears for up to four years and then reappears.

As you can see in Map 1 above, the Common Eastern Frog let may be discovered specifically around the coastlines, bushes and scrublands of the east of Australia and in Tasmania. These regions are specifically open canopy huge leaved evergreen tree cowl that is where Common Eastern Frog lets may be observed most.

CHAPTER TWO

HABITAT

Broadleaved forests are frequently called temperate rainforests and are dominated by using evergreen bushes as seen in figure 3. They are most often observed in areas with lots of cowl and which might be damp or flooded after rains such as temperate forests (Broad-leaved woodland atmosphere, 2021). Hence why the general public of the distribution of these frogs are placed in temperate forests.

Common Eastern Frog lets can also be determined in, mountains, coasts, overlaying floodplains, forests, grasslands, open and disturbed regions. Possible websites encompass transient flooded ditches, streams, ponds and dams which permit breeding to occur all year spherical (Eastern (Common) Frog let Crania signifier

CHARACTERISTICS

There are excessive distributions inside the Australian Alps region which reaches an altitude of two, 000 meters. In map 2 below, it indicates the terrain on Australia and the lighter parts imply excessive terrain and the inexperienced indicates low lying terrain.

This frog species populace should potentially lower in future years even though as with more city development occurring around coastlines and deforestation, those frogs may want to soon lower unexpectedly if we aren't careful enough

Adults Crania signifier is a small floor dwelling frog between 18-28mm lengthy. The Eastern Frog let is one maximum common and widely disbursed frogs of Eastern Australia. One distinguishing feature of the Eastern Frog let is the granular blotched black and white belly. The throat and chest of the males is white, gray (muddy white) or brown. The texture on their lower back is variable, it can either be easy with a few warts, have longitudinal ridges, or have boomerang shaped ridges over the shoulder and back. The palms and toes of the Common Frog let aren't webbed. With scattered flecks of black or gold. The tail has a rounded tip, with gold flecks on lighter colored tadpoles, but no flecks are seen at the darker variety. These tadpoles are backside dwellers feeding on microscopic particles of the substrate. They cover under the leaf litter and amongst rocks in which they are nicely camouflaged.

Tadpoles the body is quite plump and small with an stomach this is usually wider than deeper. They are light grey or brown in frame color with scattered flecks of black or gold. The tail has a rounded tip, with gold flecks on lighter colored tadpoles, however no flecks are visible on the darker range. These tadpoles are backside dwellers feeding on microscopic debris of the substrate. They disguise below the leaf litter and among rocks in which they're properly camouflaged. Eggs are very small and laid singly. Sometimes the lady lays several at a time forming a free cluster round a submerged stem or twig. More usually, the eggs are attached singly to the flowers or roll freely

LIFE CYCLE AND MATING CALL

Male Eastern Frog lets name from amongst debris and plants close to the water's edge or while floating among submerged flora. Their call can be heard all 12 months round in addition to all day long. The call is a like a cricket chirping, with three to 5 repeats "crick....Crick....Crick". This mating name is extra dominant after rain, mainly inside the cooler seasons. Hatchlings arise 7-10 days after laying. Females can first reproduce at

HABITAT AND DISTRIBUTION

The Eastern Frog let is well distributed in the course of eastern Australia. It is located in nearly all habitats from mountains to the coast, masking floodplains, forest, grasslands, open and disturbed regions. Possible sites encompass transient flooded ditches, streams, ponds and dams which allow breeding to occur all 12 months spherical. The Eastern Frog let shelters beneath logs and leaf muddle (in which it's miles nicely camouflaged), and is determined in debris near the rims of swamps and ponds.

CHAPTER THREE

IMPORTANCE AND POTENTIAL THREATS

The Eastern Frog let is not unusual and very solid in the course of Victoria with the possibility of populace sizes increasing. Possible threats to the Eastern Frog let include: urbanization and tourism, lack of brief swimming pools, and habitat fragmentation. These frogs have a excessive fecundity price, laying as much as 51-200 eggs/woman/yr. Throughout Australia, the Common Frog let covers a variety of eighty,001-a million km² with an expected general populace of >50000 adults.

BREEDING SEASON

For the duration of the yr. Females lay a hundred-150 pigmented eggs commonly found singly or in small organizations frequently attached to submerged vegetation.

Eggs are very small and laid singly. Sometimes the girl lays several at a time forming a unfastened cluster around a submerged stem or twig. More generally, the eggs are attached singly to the flowers or roll freely.

WHAT YOU COULD DO TO HELP!

Sometimes people can be a frog's worst enemy! Although the Common Frog let stays abundant and good sized, lack of habitat because of clearing of land and improvement threatens some species of frogs. These species need your assist. Frogs are at risk of pollution, so maintain waterways clean via now not dumping waste or toxic drinks into your drains and creeks. Frogs are also susceptible to an infectious disorder caused by the cheered fungus that could cause them to sick or kill them. Help shield

Common Frog lets by way of not touching or shifting them from one vicinity to every other. Create a frog-pleasant outside with the aid of composting and keep away from using chemical substances like herbicides. Providing rocks, logs, leaf clutter and appropriate shrubs round your private home will keep the frogs glad. In go back, you'll be extremely joyful with the sound of their calls.

DISTRIBUTION

The common frog let stages from South Japanese Australia, from Adelaide to Melbourne, up the jape coast to Brisbane. It additionally inhabits a majority of

Tasmania. It is one of the most normally encountered frog species within its Variety, due to its capacity to occupy numerous habitat types.

CHAPTER FOUR

DESCRIPTION

The commonplace Japanese frog let is a small frog (3 centimeters), of brown or gray coloration of numerous sunglasses. The frog is of extremely variable markings, with super variety generally observed within limited populations. A darkish, triangular mark is observed at the upper lip, with darker bands on the legs. A small white spot is on the bottom of every arm. The dorsal and ventral surfaces are very variable. The dorsal surface may be easy, warty or have longitudinal skin folds. The coloration varies from darkish brown, fawn, light and dark grey. The coloration of the ventral floor is similar to the dorsal floor, however mottled with white spots.

ECOLOGY AND BEHAVIOUR

The commonplace Japanese frog let will call within a large refrain of males close to a still water supply, or slow flowing creek. The call of the male is a crik-crik-crik, this is heard all yr round, throughout moist and dry conditions. An common of about 2 hundred eggs are laid in small clusters attached to submerged flowers, the tadpoles and eggs continue to exist in 14–15

°C water. Tadpoles are commonly brown and attain approximately 36mm in period. Development is rather short; however it is dependent on environmental conditions. At a temperature of 15 °C development can range from 6 weeks to extra than three months. Metamorphic frogs are very small, approximately 8 mm.

The food plan of the species includes small insects, a whole lot smaller in contrast to their length than maximum frogs.

DISTINGUISHING CHARACTERISTICS

• Commonly harassed with pickerel frog, with which it shares comparable color and patterning

• Medium-sized, approximately 2 to a few. Five inches in duration

• Usually inexperienced (occasionally brown) above with irregularly organized rounded dark spots with skinny light borders on again between outstanding ridges (dorsolateral folds) along sides of again from at the back of eye to near groin

•   Underside white to grayish-white

•   Breeding call is an extended low snore with individual notes generally discernible, interspersed with brief grunts and chuckles

DIET

•   Tadpoles eat algae, and adults devour a extensive sort of invertebrates, in particular beetles

Water

1.   The water portion of the enclosure includes aquarium gravel. Gravel may be sloped to permit entrance or exit to the water supply. 2. Chlorine- and

chloramines-free water need to be used. Allowing water to sit for 24 hours will assist take away chlorine and an amphibian-safe dechlorinator can be used to put off chloramines. 3. Filtration is non-compulsory. Constant water vibrations are believed to cause sensory overload, so a smooth waft version would be excellent if a filter out is used. Four. Water must be modified often, without or with a filter out. If no filter out is used, the water must be modified day by day.

THE END

Printed in Great Britain
by Amazon

23559141R00020